Attitude and Energy

101 Meeting Topics

Notes

Notes

Work Book

Acknowledgments

The Keys nobody had time to Teach: Edited Version

The Seed of a Woman "Baby or Offspring" Edited Version

Rule And Guide

Faith And Practice

The GOAT With Common Sense

The King

All Books listed above are on Amazon and Available in all bookstores Now.

1. *Mondays belong to the Go-Getters*

2. *If you cannot do great things, do small things in a Great Way.*

3. _**Have a Great Day.**_

4. *The Journey of a Thousand miles begins with one Step.*

5. _**The Best Preparation for Tomorrow is Doing your best Today.**_

6. _**You have to fight through some bad days to Earn the best days of your Life.**_

7. *If you Truly love someone, Being Faithful is Easy.*

8. _**Every Morning you have two Choices: Continue to Sleep with your Dreams or Wake up and Chase them.**_

9. *Sometimes its not the People who change, it's the mask that Falls off.*

_10._   **_The Fact that you aren't where you want to be should be Motivation Enough._**

11. *Nothing is Impossible. The world itself says 'I'm Possible.*

12. *All Our Dreams Can Come true if We Have the Courage to Pursue them.*

13.   The only person you should try to be better than is the Person you were Yesterday.

14. *Spread Love Everywhere you Go.*

15. *<u>Success is not Final, Failure is not Fatal.</u>*

16. *If you want to Go fast, Go Alone. If you want to Go far, Go Together.*

17. *Happiness is an Inside Job.*

18. Do What makes you Happy because You'll be criticized anyway.

19. _Do not set yourself on Fire in order to keep others Warm._

_20._   **_Opinions aren't Facts. Stop worrying about what people think about you._**

21. *Love all. Trust Few. Do Wrong to None.*

22. *<u>If you want it, Go Get It.</u>*

23. *The Climb is Tough, But the View from the Top is Worth it.*

24. *Loyalty is Rare. If you find it, Keep it.*

25. *<u>The only person Responsible for your success is You.</u>*

26. *If you Look at what you have in Life, You will Always have More.*

27. *Something very Special is coming your way. Are you Ready For It?*

28. *Believe in Yourself.*

29. *Take it Day by Day. Don't Stress Too Much About Tomorrow.*

30. *My Mission in Life is not Merely to survive, But to Thrive.*

31. *Be Nice to Yourself, Always.*

32. *<u>Hate is Heavy, Let it Go.</u>*

33. *You Must Do the Things you think you Cannot Do.*

34. Don't Wait. The Time will never be Just Right.

35. *<u>Winners Never Quit. Quitters Never Win.</u>*

36. *You Are Never Too Old to Set another Goal or to Dream A New Dream.*

37. *Life is Like a Bicycle. To Keep your Balance, You Must keep Moving.*

38. _Stay Close TO anything That Makes you Glad You are Alive._

39. <u>No Excuses.</u>

_40._   **_Be Humble. Be Hungry. And always be the Hardest worker in the Room._**

41. _Failure is a Success if We Learn From it._

42.   Work until you No long have to Introduce yourself.

43. *<u>You Have to Believe in Yourself when No one Else does. That's what Makes you a Winner.</u>*

44. *Keep Going. In a year from now, you'll thank yourself.*

45. *Wake up with determination. Go to bed with Satisfaction.*

46. *You get what you focus on, so focus on what you want.*

47. *It's your life. Don't let anyone make you feel guilty for living it your way.*

48._If you wait for the perfect conditions, you'll never get anything done._

_49._   **_I may not be there yet, but I'm closer than I was yesterday._**

50. <u>Difficult roads often lead to beautiful destinations.</u>

51. *<u>No one is you, and that is your super power.</u>*

52. *The Greatest adventure you can take is to live the life of your dreams.*

53. *The best time for new beginnings is right now.*

54. <u>*One day or day one. It's your decision.*</u>

55. *_Success doesn't just come and find you, you have to go out and get it._*

56. *<u>If you can dream it, you can achieve it.</u>*

57.   You cannot have a positive life with a negative mind.

58. _Every day may not be good, but there is something good in every day._

59. _Slow progress is still progress._

60. *You can't go back and change the beginning, but you can start where you are and change the ending.*

61. *<u>If it comes, let it. If it goes, let it.</u>*

62. _You were born an original, don't die a copy._

63. *The days that break you are the days that make you.*

64. *Fight for your dreams.*

65. _Sometimes the smallest step in the right direction can end up being the biggest step in your life._

66. _Someone will always be prettier. Someone will always be smarter. Someone will always be younger. But the will never be you._

67. *Not everyone you lose is a loss.*

68. *<u>Mistakes are proof that you're trying.</u>*

69. *Real friends are a blessing.*

70. *<u>Don't stop until you're proud.</u>*

71. _One day you'll laugh at how much you let this matter._

72.   <u>Never mistake silence for ignorance, calmness for acceptance, or kindness for weakness.</u>

_73._   **_The only way to win with a toxic person is to not play._**

74. _When you fail, that is when you get a step closer to success._

75. *<u>You can if you think you can.</u>*

76. _Your time is way too valuable to be wasting it on people who can't accept who you are._

77. *Don't doubt yourself, that's what haters are for.*

78.   It is amazing what you can accomplish if you do not care who gets credit.

79. *<u>Tough time never last, but tough people do.</u>*

80.   Hard work beats talent when talent doesn't work hard.

81. *You can be as great as you want to be. If you believe in yourself and have the courage, it can be done.*

82. *Every great story happened when someone decided not to give up and kept going no matter what.*

83. *Don't stop when you're tired, stop when you're done.*

84.   Action is the foundational key to all success.

85. _**Dream big. Stop thinking small.**_

86. *Find someone who knows you're not perfect but treats you as if you are.*

87.   Never announce your moves before you make them.

88. _I respect those who tell me the truth, no matter how hard it is._

89. <u>Don't decrease the goal. Increase the effort.</u>

90. *When you learn to survive alone, you can survive anything.*

91. *<u>Pain is temporary. Quitting last forever.</u>*

92. *Mindset is what separates the best from the rest.*

93. *__Stop expecting loyalty from people who can't give you honesty.__*

94. *The moment you start acting like life is a blessing it starts feeling like one.*

95. _Keep away from people who try to belittle your ambition._

96. _**It's never too late for a new beginning.**_

97. *<u>Impossible is just an option.</u>*

98. *<u>Just be yourself.</u>*

99. *You are what you do, not what you say you'll do.*

100. *__Thank you for the pain, it made me raise my game.__*

_101._   **_Patience is power._**

Notes

Notes

Notes

Notes

Contact Us

WestEastJackson@gmail.com

@WestEastJackson

@teamloveiv

Team Love, LLC

Aldie, Va

Loudoun County

Notes

Notes

www.ingramcontent.com/pod-product-compliance
Lightning Source LLC
Chambersburg PA
CBHW071530220526
45469CB00003B/715